Harnessing Water Power A Comprehensive Introduction to Hydroelectric Engineering for Students

AF487408

Hemingway

Title: Harnessing Water Power A Comprehensive Introduction to Hydroelectric Engineering for Students

This book is a self-published work by the author Hemingway

ISBN:

TABLE OF CONTENTS

Chapter 9: Case Studies in Hydroelectric Engineering   58

Chapter 10: Conclusion   64

Chapter 1: Introduction to Hydroelectric Energy Engineering

The Importance of Hydroelectric Power

Hydroelectric power is a topic of great importance in the field of energy engineering. As students in this discipline, it is crucial to understand the significance of hydroelectric power and its impact on our society and the environment. In this subchapter, we will delve into the various aspects that make hydroelectric power an indispensable source of renewable energy.

First and foremost, hydroelectric power is a clean and renewable source of energy. Unlike fossil fuels, which contribute to air pollution and global warming, hydroelectric power generation does not emit harmful greenhouse gases. It relies on the force of moving water, which is constantly replenished by the water cycle, making it a sustainable and eco-friendly solution to our energy needs.

Furthermore, hydroelectric power plays a vital role in meeting the increasing demand for electricity worldwide. As the global population continues to grow, so does the need for reliable and affordable energy. Hydroelectric power plants can generate a significant amount of electricity, making it a feasible option for countries striving to meet their growing energy demands.

In addition to providing electricity, hydroelectric power also offers multiple benefits in terms of water management. Hydroelectric dams serve as reservoirs, allowing for the controlled storage and release of water. This enables us to manage water resources effectively, ensuring a stable water supply for agricultural, industrial, and domestic use. Moreover, hydroelectric power plants can assist in flood control by regulating water levels during periods of heavy rainfall, mitigating the risk of destructive floods.

Another crucial aspect of hydroelectric power is its ability to support local economies. The construction and operation of hydroelectric power plants create job opportunities, thus stimulating economic growth in the surrounding areas. Additionally, hydroelectric power plants often become tourist attractions, attracting visitors and boosting the tourism industry, which further contributes to the local economy.

Understanding the importance of hydroelectric power is essential for energy engineering students. As future professionals in this field, it is our responsibility to explore and develop sustainable energy solutions that can meet the demands of our growing world population. By harnessing the power of water, we can make a significant difference in reducing our reliance on fossil fuels and combating climate change.

In conclusion, hydroelectric power is a crucial component of the energy engineering discipline. Its clean and renewable nature, ability to meet growing energy demands, water management benefits, and economic contributions make it an indispensable source of power. As students, it is vital for us to recognize the importance of hydroelectric power and its potential to shape a sustainable future for our planet.

Overview of Hydroelectric Power Plants

Hydroelectric power plants play a crucial role in the field of energy engineering as sustainable and renewable sources of electricity. Harnessing the power of water, these plants have been instrumental in meeting the growing global energy demands while minimizing the environmental impact associated with traditional fossil fuel-based power generation. In this subchapter, we will delve into the various aspects of hydroelectric power plants, their components, and their significance in the field of energy engineering.

To begin with, a hydroelectric power plant is a facility designed to convert the kinetic and potential energy of flowing water into electrical energy. The process involves the use of a dam to create a reservoir, which stores water at an elevated level. This stored water is then released through turbines, causing them to spin and generating electricity through connected generators. The amount of electricity produced depends on the volume of water flowing through the turbines and the height from which it falls, known as the head.

There are different types of hydroelectric power plants, including conventional and pumped storage plants. Conventional plants utilize natural water flow to generate electricity, while pumped storage plants store excess electricity during low-demand periods by pumping water from a lower reservoir to a higher one. This stored water is then released during peak demand periods to generate additional electricity.

The components of a hydroelectric power plant include the dam, reservoir, intake structure, penstock, turbine, generator, and transmission lines. Each component plays a vital role in the overall functioning of the plant, ensuring efficient conversion of water energy into electrical energy.

Hydroelectric power plants offer numerous advantages. They provide a clean and renewable source of energy, reducing greenhouse gas emissions and helping combat climate change. Additionally, they offer flexibility in electricity generation, as the water flow can be controlled to meet fluctuating demands. Furthermore, hydroelectric power plants contribute to water resource management by regulating water flow and preventing floods.

In conclusion, hydroelectric power plants are an essential part of the energy engineering field, offering sustainable and efficient electricity generation. Understanding their components, working principles, and advantages is crucial for students studying energy engineering. By harnessing the power of water, these plants contribute to a greener future while meeting the increasing global energy demands.

Historical Development of Hydroelectric Engineering

Hydroelectric power has been harnessed for centuries as a reliable and sustainable source of energy. Understanding the historical development of hydroelectric engineering is crucial for students in the field of energy engineering. This subchapter delves into the fascinating journey of harnessing water power and highlights significant milestones that have shaped the field.

The roots of hydroelectric engineering can be traced back to ancient civilizations. The Greeks and Romans were among the first to use waterwheels to grind grains and saw timber. However, it was not until the late 19th century that hydroelectric power started to gain traction as a viable source of electricity.

The industrial revolution played a pivotal role in the development of hydroelectric engineering. In 1878, the world's first hydroelectric power plant was installed on the Fox River in Wisconsin, USA. This marked the beginning of a new era. Soon after, Niagara Falls became the site of the world's first large-scale hydroelectric power plant, generating electricity for both industrial and residential use.

As hydroelectric power gained popularity, engineers faced various challenges. One significant obstacle was the transmission of electricity over long distances. The invention of alternating current (AC) by Nikola Tesla and George Westinghouse in the late 19th century revolutionized the field, as it allowed for efficient long-distance power transmission. This breakthrough paved the way for the construction of larger hydroelectric power plants in remote areas, utilizing the vast potential of rivers and dams.

In the early 20th century, hydroelectric engineering witnessed further advancements. The construction of the Hoover Dam in the United States, completed in 1935, was a monumental achievement. It not only provided electricity to millions but also revolutionized irrigation and water supply management in the region.

In recent times, hydroelectric engineering has embraced new technologies and methodologies. The advent of computer-aided design and simulation tools has enhanced the efficiency and accuracy of plant designs. Additionally, the focus has shifted towards incorporating sustainable practices, such as fish-friendly turbine designs and environmentally conscious dam operations.

Understanding the historical development of hydroelectric engineering allows students to appreciate the progress made in this field. It also provides insights into the challenges that engineers faced and the solutions they devised. By studying the past, students can gain valuable knowledge that can be applied to future innovations in the field of energy engineering.

In conclusion, the historical development of hydroelectric engineering showcases the continuous evolution of this field. From ancient waterwheels to modern-day mega-dams, engineers have constantly pushed the boundaries to harness the power of water. As students, understanding this journey helps us appreciate the significance of hydroelectric power and motivates us to create a sustainable future through innovative engineering solutions.

Chapter 2: Fundamentals of Water Power

Water as a Renewable Energy Source

In recent years, the need for sustainable and renewable energy sources has become increasingly crucial. As the world grapples with the challenges of climate change and dwindling fossil fuel reserves, it is imperative for students in the field of Energy Engineering to explore alternative solutions to meet the growing energy demand. One such solution that holds immense potential is harnessing water power as a renewable energy source.

Water, in its various forms, has been utilized for centuries to generate power. From ancient water wheels to modern hydroelectric dams, humans have tapped into the immense power of water to produce electricity. This subchapter aims to provide a comprehensive introduction to hydroelectric engineering, helping students understand the principles, technology, and environmental impacts associated with water as a renewable energy source.

One of the primary advantages of water as an energy source is its renewability. Unlike fossil fuels that are finite and non-renewable, water is an abundant resource that can be replenished through natural processes such as rainfall and evaporation. This makes it an attractive option for long-term and sustainable energy production.

Hydroelectric power plants, which convert the kinetic energy of flowing or falling water into electricity, are the most common form of harnessing water power. This subchapter will delve into the various components and processes involved in hydroelectric engineering, including dams, reservoirs, turbines, and generators. Students will gain insight into the mechanics of hydroelectric power generation, from the capturing and storing of water to the conversion of mechanical energy into electrical energy.

Additionally, the subchapter will explore the environmental impacts associated with hydroelectric power. While water is a renewable resource, the construction and operation of hydroelectric plants can have significant ecological consequences. Students will learn about the potential effects on aquatic ecosystems, fish migration patterns, and the displacement of local communities. Balancing the benefits and drawbacks of hydroelectric power is vital in ensuring sustainable energy practices.

Understanding water as a renewable energy source is crucial for students in the field of Energy Engineering. By exploring the principles and technology behind hydroelectric engineering, students can contribute to the development of sustainable energy solutions. As the world continues to shift towards renewable sources, the knowledge gained from this subchapter will empower students to make informed decisions and lead the way in creating a greener future.

Basic Principles of Hydropower Generation

Hydropower, also known as water power, is the most widely used form of renewable energy worldwide. It involves harnessing the energy from flowing or falling water to generate electricity. This subchapter will introduce you to the basic principles of hydropower generation, providing a comprehensive overview of this exciting field within the realm of energy engineering.

To understand hydropower generation, it is important to grasp the fundamental concept of energy conversion. In simple terms, hydropower works by converting the potential energy of water into mechanical energy, which is then transformed into electrical energy. This conversion process primarily involves three key components: dams, turbines, and generators.

Dams play a crucial role in hydropower generation as they control the flow of water. By creating a reservoir, dams ensure a steady supply of water for the power plant. As water is released from the reservoir, it flows through pipes or penstocks, which are connected to turbines.

Turbines are the heart of a hydropower plant. They receive the kinetic energy of flowing water and convert it into mechanical energy. Turbines can be classified into various types, including Pelton, Francis, and Kaplan turbines, each designed for specific flow conditions. The choice of turbine depends on factors such as water velocity, head, and power requirements.

Once the mechanical energy is generated, it is transmitted to generators. Generators consist of a rotating shaft and electromagnets. As the shaft rotates, it creates a magnetic field, inducing an electric current within the generator's coils. This current is then conducted through transmission lines, delivering electricity to homes, businesses, and industries.

Hydropower boasts numerous advantages. It is renewable, reliable, and produces virtually no greenhouse gas emissions. It also provides flexibility in terms of controlling electricity generation, allowing for peak load management and storage of excess energy in the form of pumped storage.

However, hydropower generation also faces certain challenges. The construction of dams can have significant environmental impacts, affecting aquatic ecosystems and displacing communities. Therefore, sustainable hydropower development and careful environmental planning are crucial to minimize these negative effects.

In conclusion, understanding the basic principles of hydropower generation is essential for students in the field of energy engineering. By harnessing the power of flowing or falling water, hydropower offers a clean and reliable source of energy. As the world seeks sustainable solutions to combat climate change, hydropower continues to play a vital role in the global energy mix.

Types of Hydroelectric Turbines

In the world of hydroelectric engineering, turbines play a vital role in harnessing the power of water and converting it into electricity. These ingenious machines are designed to capture the kinetic energy of flowing or falling water and convert it into mechanical energy, which is then used to generate electrical power. There are various types of hydroelectric turbines, each with their unique characteristics and applications. In this subchapter, we will explore some of the most commonly used types of hydroelectric turbines.

1. Impulse Turbines: Impulse turbines are primarily used in high-head installations, where water flows with high velocity through nozzles or jets. These turbines work on the principle of Newton's third law of motion, where the change in momentum of the water jet creates a force that turns the turbine's runner. Pelton turbines are a popular example of impulse turbines and are known for their high efficiency at high heads.

2. Reaction Turbines: Unlike impulse turbines, reaction turbines are used in low-head installations, where water flows under pressure. These turbines work on the principle of Newton's second law of motion, where the water's pressure energy is converted into kinetic energy as it passes through the runner. Francis turbines are the most widely used reaction turbines, known for their versatility in handling a wide range of flow rates and heads.

3. Kaplan Turbines: Kaplan turbines are a variation of the reaction turbine designed specifically for low-head installations. These turbines have adjustable blades that allow them to optimize their performance at different flow rates. Kaplan turbines are commonly used in river-

based hydroelectric power plants, where the water flow varies significantly throughout the year.

4. Turgo Turbines: Turgo turbines are impulse turbines that operate at medium to high heads. These turbines feature a unique design with a runner that resembles a water wheel and buckets shaped like spoons. The water hits the buckets at an angle, creating a force that turns the runner. Turgo turbines are widely used in small-scale hydroelectric power plants due to their simplicity and efficiency.

5. Crossflow Turbines: Crossflow turbines, also known as Banki-Michell turbines, are a type of reaction turbine designed for low to medium-head installations. These turbines have a vertical shaft and a runner with blades curved in a helical pattern. Crossflow turbines are known for their compact size and ability to operate efficiently even at low flow rates.

Understanding the different types of hydroelectric turbines is crucial for energy engineering students. Each turbine has its advantages and limitations, making it suitable for specific hydroelectric projects. By selecting the appropriate turbine for a given site and water conditions, engineers can ensure maximum efficiency and sustainable energy production.

In conclusion, hydroelectric turbines are the heart of any hydroelectric power plant. The various types of turbines, such as impulse turbines like Pelton turbines, reaction turbines like Francis turbines, and specialized turbines like Kaplan turbines, Turgo turbines, and crossflow turbines, offer flexibility and efficiency in harnessing water power. By studying and understanding these turbine types, students in the field of energy engineering can contribute to the development of sustainable and clean energy solutions for the future.

Chapter 3: Hydrological Considerations in Hydroelectric Engineering

Understanding Water Resources

Water is a vital resource that plays a crucial role in our everyday lives. From quenching our thirst to powering our homes, water resources are the backbone of numerous industries, including energy engineering. In this subchapter, we will delve into the intricacies of water resources and their significance in the field of hydroelectric engineering.

To begin with, it is essential to understand the concept of water resources. Water resources refer to the freshwater sources available on our planet, such as rivers, lakes, and underground aquifers. These resources are not only essential for sustaining life but also for generating electricity through the process of hydroelectric power generation.

The availability and quality of water resources greatly impact the efficiency and sustainability of energy engineering projects. Hydroelectric power plants rely on the force of flowing or falling water to generate electricity. Therefore, the quantity and reliability of water resources in a given area determine the feasibility of constructing a hydroelectric power plant.

Furthermore, understanding the hydrological cycle is crucial in comprehending water resources. The hydrological cycle involves the continuous movement of water through various stages, including evaporation, condensation, precipitation, and runoff. This cycle replenishes water resources and determines their availability at different times and locations.

Water resources also face various challenges, including climate change, pollution, and overexploitation. Climate change affects water resources by altering precipitation patterns and exacerbating droughts and floods. Pollution, caused by human activities, poses a significant threat to water quality and ecosystem health. Overexploitation of water resources, especially in arid regions, can lead to water scarcity and ecological imbalances.

In the context of energy engineering, students need to understand the sustainable management and conservation of water resources. This involves employing efficient water use strategies, investing in wastewater treatment technologies, and promoting responsible water consumption practices.

In conclusion, understanding water resources is paramount for students in the field of energy engineering. It is essential to grasp the significance of water as a valuable resource and its role in hydroelectric power generation. By comprehending the hydrological cycle, challenges, and sustainable management practices, students can contribute to the efficient and responsible utilization of water resources in the pursuit of clean and renewable energy.

Estimating Water Availability and Flow Rates

Water availability and flow rates are crucial factors to consider when designing and implementing hydroelectric projects. Understanding how to estimate these variables is essential for energy engineering students studying hydroelectric engineering. This subchapter will provide a comprehensive introduction to estimating water availability and flow rates, equipping students with the knowledge and tools to make informed decisions in their future careers.

To accurately estimate water availability, various factors must be considered. These include precipitation patterns, watershed characteristics, and evaporation rates. Students will learn how to analyze historical data, such as rainfall records and streamflow measurements, to determine the average water availability in a specific region. They will also explore methods for predicting future water availability, considering climate change and potential shifts in precipitation patterns.

Flow rates, on the other hand, refer to the amount of water passing through a hydroelectric facility at a given time. Students will delve into the concept of flow rate estimation, covering techniques such as stream gauging and flow measurement devices. They will learn about the different flow regimes, including steady flow and unsteady flow, and how to calculate flow rates under varying conditions.

Additionally, this subchapter will introduce students to the concept of flow duration curves. Flow duration curves provide valuable insights into the distribution of flow rates over a specific period, helping engineers determine the feasibility and efficiency of a hydroelectric project. Students will learn how to construct and interpret flow duration curves, enabling them to make accurate predictions about the availability of water at different flow rates.

Furthermore, students will explore the impact of climate change on water availability and flow rates. They will study models and methodologies used to assess potential changes in precipitation patterns and the resulting implications for hydroelectric engineering. Understanding these future scenarios is crucial for designing resilient and sustainable hydroelectric projects.

By the end of this subchapter, students will have a solid understanding of how to estimate water availability and flow rates for hydroelectric projects. This knowledge will equip them with the skills necessary to design and implement efficient and sustainable energy solutions. Whether working in the field of energy engineering or pursuing further research in hydroelectric engineering, this subchapter serves as an invaluable resource for students seeking to harness the power of water for renewable energy generation.

Hydrological Data Collection and Analysis

Understanding the hydrological cycle and accurately assessing water resources are crucial aspects of hydroelectric engineering. The subchapter "Hydrological Data Collection and Analysis" delves into the methods and tools used to measure and analyze hydrological data, providing students in the field of Energy Engineering with a comprehensive introduction to this essential aspect of the industry.

The subchapter begins by highlighting the significance of hydrological data in the planning, design, and operation of hydroelectric power projects. It emphasizes the need to collect accurate and reliable data to assess the availability, quantity, and quality of water resources. Students will learn about the various parameters that must be measured, such as precipitation, river flows, groundwater levels, and water quality indicators.

The subchapter then explores the different techniques and instruments employed for hydrological data collection. It covers traditional methods such as rain gauges, streamflow gauges, and groundwater wells, as well as modern technologies like remote sensing, satellite imagery, and computer modeling. Students will gain an understanding of the advantages and limitations of each method, allowing them to choose the most appropriate techniques for specific projects.

Moreover, the subchapter provides a detailed overview of the process of data analysis. Students will learn how to organize and manage collected data using spreadsheets and databases. They will be introduced to statistical methods for analyzing and interpreting hydrological data, including frequency analysis, trend analysis, and correlation analysis. Real-world examples and case studies will be included to illustrate the practical applications of these analytical techniques.

To ensure that students grasp the concepts effectively, the subchapter offers interactive exercises and practical assignments. These activities will encourage hands-on learning, allowing students to apply their knowledge to real hydrological data sets and develop valuable skills in data collection, analysis, and interpretation.

By the end of the subchapter, students will have a solid foundation in hydrological data collection and analysis. They will understand the importance of accurate data in hydroelectric engineering and be equipped with the necessary tools and techniques to assess water resources effectively. This knowledge will empower them to make informed decisions and contribute to the sustainable development of hydropower projects in the future.

In summary, "Hydrological Data Collection and Analysis" is an essential subchapter in the book "Harnessing Water Power: A Comprehensive Introduction to Hydroelectric Engineering for Students." It provides students in the niche of Energy Engineering with a comprehensive understanding of the methods, tools, and techniques involved in collecting and analyzing hydrological data, enabling them to make informed decisions in the field of hydropower.

Chapter 4: Site Selection and Design of Hydroelectric Power Plants

Factors Influencing Site Selection

When it comes to harnessing water power for hydroelectric engineering, selecting the right site is of utmost importance. There are various factors that need to be considered in order to ensure the success and efficiency of a hydroelectric project. This subchapter will explore the key factors that influence site selection in the field of energy engineering.

1. Water Availability: One of the primary factors to consider is the availability of water. Sufficient water supply is essential for generating a consistent and reliable amount of power. The quantity and quality of water must be carefully assessed to ensure that it meets the project's requirements.

2. Topography: The topography of the site plays a crucial role in determining the feasibility of a hydroelectric project. The presence of steep slopes, valleys, or canyons can provide the necessary elevation changes required for the construction of dams and powerhouses. Additionally, the shape and size of the reservoir can affect the overall efficiency of power generation.

3. Geological Conditions: The geological characteristics of a site can significantly impact its suitability for hydroelectric projects. The stability of the ground, rock formations, and soil composition must be thoroughly examined to avoid potential risks such as landslides or erosion that could compromise the infrastructure.

4. Environmental Considerations: As responsible energy engineers, it is important to consider the environmental impact of hydroelectric projects. The chosen site should minimize disruption to natural habitats, protect endangered species, and maintain the ecological balance. Environmental assessments and studies should be conducted to identify potential risks and develop mitigation measures.

5. Proximity to Transmission Lines: The proximity of the site to existing transmission lines is an important consideration in reducing the cost and complexity of power distribution. Identifying sites close to transmission infrastructure can streamline the process of connecting the hydroelectric plant to the power grid.

6. Socioeconomic Factors: Socioeconomic factors, such as population density, land use patterns, and proximity to communities, need to be taken into account. A thorough analysis should be conducted to evaluate the potential social and economic impacts of the project on local communities, livelihoods, and cultural heritage.

In conclusion, selecting an appropriate site for hydroelectric projects involves a comprehensive evaluation of water availability, topography, geological conditions, environmental considerations, proximity to transmission lines, and socioeconomic factors. By carefully considering these factors, energy engineers can ensure the successful implementation of hydroelectric projects that are both economically viable and environmentally sustainable.

Environmental Impacts and Mitigation Measures

In the field of hydroelectric engineering, it is crucial to consider the environmental impacts associated with harnessing water power for energy generation. While hydroelectric power is a clean and renewable source of energy, it is not without its own set of challenges. This subchapter aims to explore the various environmental impacts of hydroelectric projects and provide potential mitigation measures to minimize their adverse effects.

One of the primary environmental concerns associated with hydroelectric power is the alteration of aquatic ecosystems. The construction of dams and reservoirs can disrupt the natural flow of rivers, leading to changes in water temperature, sediment transport, and fish migration patterns. To mitigate these effects, engineers can incorporate fish ladders or fish bypass systems to facilitate fish passage and maintain their natural migration routes. Additionally, conducting thorough environmental impact assessments before project implementation can help identify sensitive areas and develop appropriate mitigation strategies.

Another significant impact is the loss of habitat due to the flooding caused by reservoirs. This can result in the displacement of wildlife and vegetation. To mitigate habitat loss, engineers can adopt a strategic approach of reservoir zoning, which involves the creation of protected areas and the restoration of degraded habitats. This ensures the preservation of biodiversity and allows for the sustainable coexistence of hydroelectric projects and ecosystems.

Sedimentation is yet another environmental challenge in hydroelectric engineering. The construction of dams can trap sediments, leading to downstream erosion and the loss of fertile soil. To address this issue, engineers can incorporate sediment bypass systems or develop sediment management strategies to minimize the negative impacts on downstream ecosystems and agricultural lands.

Furthermore, hydroelectric projects may have social impacts, such as the displacement of local communities or the alteration of traditional livelihoods. Adequate stakeholder engagement and community participation are essential in the planning and implementation stages to ensure that the concerns of affected communities are addressed. This can include providing compensation, improving infrastructure, or offering alternative livelihood options.

In conclusion, while hydroelectric power offers numerous benefits, it is essential to address the environmental impacts associated with its generation. By implementing appropriate mitigation measures, such as fish passage systems, habitat restoration, sediment management, and community engagement, hydroelectric engineering can strike a balance between energy generation and environmental sustainability. As future energy engineers, it is crucial to be aware of these impacts and work towards developing innovative solutions for a sustainable and harmonious coexistence of hydroelectric projects and the surrounding ecosystems.

Design Considerations for Dam and Reservoir

When it comes to harnessing water power, the design of dams and reservoirs plays a crucial role in the success of a hydroelectric project. In this subchapter, we will explore the key design considerations for dams and reservoirs, providing students in the field of energy engineering with a comprehensive understanding of this critical aspect of hydroelectric engineering.

First and foremost, the choice of dam type is a fundamental decision in the design process. There are various types of dams, including arch dams, gravity dams, buttress dams, and embankment dams. Each type has its advantages and disadvantages, and the selection depends on factors such as site conditions, available materials, and project requirements. Students must thoroughly study each type to determine the most suitable option for a given project.

Another crucial consideration is the stability and safety of the dam structure. Dams must be designed to withstand the forces exerted by the water in the reservoir, including hydrostatic pressure, wave action, and seismic events. Students should learn about the principles of structural analysis and design, including factors such as load distribution, material strength, and the importance of proper foundation design.

Reservoir management is equally important in the design process. The size and shape of the reservoir should be carefully determined by considering factors such as water availability, storage capacity, and environmental impact. Students should also learn about sedimentation management techniques to prevent the accumulation of sediments in the reservoir, which can reduce storage capacity and affect the performance of the hydroelectric plant.

Furthermore, the ecological impact of dam construction should not be overlooked. Dams can significantly alter aquatic ecosystems, affecting fish migration, water quality, and the overall biodiversity of the area. Students should be aware of the environmental regulations and mitigation measures in place to minimize these impacts, such as fish ladders, fish bypass systems, and environmental flow releases.

Lastly, the design of a dam and reservoir should incorporate considerations for future maintenance and potential modifications. Regular inspections, maintenance activities, and the ability to adapt to changing conditions are essential for the sustainable operation of a hydroelectric project.

In conclusion, the design considerations for dams and reservoirs are of paramount importance in hydroelectric engineering. Students specializing in energy engineering must understand the various dam types, structural design principles, reservoir management techniques, ecological impacts, and maintenance requirements to ensure the successful implementation of hydroelectric projects. By mastering these concepts, students will be equipped to contribute to the sustainable development of the energy industry and the effective utilization of water resources.

Design Considerations for Powerhouse and Turbines

In the world of energy engineering, the design of powerhouses and turbines plays a crucial role in the efficient harnessing of water power. Powerhouses serve as the heart of any hydroelectric project, housing the turbines and other essential equipment. Therefore, it is imperative for students in the field of energy engineering to grasp the key design considerations for powerhouses and turbines.

First and foremost, the location of the powerhouse is a critical factor. It should be strategically placed to minimize transmission losses and to ensure the availability of a sufficient water supply. Topographical surveys and hydraulic studies are crucial to determine the optimal location, taking into account factors such as elevation, water flow, and accessibility.

The size and capacity of the powerhouse should align with the desired energy output. This requires careful consideration of the water flow rate and head available at the site. The design must also account for any potential future expansion, allowing for the installation of additional turbines if required.

The selection of turbines is another important aspect of the design process. The choice of turbine depends on the specific site conditions, including the head and flow rate of water. Francis, Kaplan, and Pelton turbines are commonly used in hydroelectric power plants, each with its own advantages and limitations. Students must familiarize themselves with the characteristics and performance of these turbines to make informed design decisions.

Efficiency is a key consideration in hydroelectric power generation. The design should aim to maximize the conversion of water energy into electrical energy. This can be achieved through optimizing the turbine design, ensuring a smooth flow of water through the turbine blades, and minimizing losses due to friction and turbulence.

Additionally, safety measures should be incorporated into the design of powerhouses and turbines. This includes proper ventilation systems, fire suppression systems, and structural integrity considerations to withstand potential natural disasters such as earthquakes and floods.

The integration of automation and control systems is also vital in modern hydroelectric power plants. Students should understand the principles of turbine control, monitoring systems, and data analysis to ensure efficient operation and maintenance.

In conclusion, the design considerations for powerhouses and turbines in hydroelectric engineering are multifaceted and require a comprehensive understanding of various factors. By considering location, size, turbine selection, efficiency, safety, and control systems, students can contribute to the development of sustainable and efficient hydroelectric power plants.

Chapter 5: Components of a Hydroelectric Power Plant

Dam and Reservoir Systems

Introduction:

In the realm of energy engineering, one of the most significant sources of renewable energy is hydroelectric power. Harnessing the force of flowing water to generate electricity has been a game-changer in the field of sustainable energy production. At the heart of this process lies the dam and reservoir system, a critical component in hydroelectric engineering. In this subchapter, we will explore the functionalities, design principles, and benefits of dam and reservoir systems.

The Role of Dams:

Dams play a pivotal role in hydroelectric power generation. These massive structures are built across rivers to create artificial lakes, known as reservoirs. The primary purpose of a dam is to store water, forming a reservoir that can release controlled amounts of water as needed. This control allows for the regulation of water flow, ensuring a steady supply of water to the power generation facilities.

Design and Construction:

The design of dams and reservoir systems requires careful consideration of various factors, such as topography, geology, and environmental impact. Engineers employ advanced techniques to select the most suitable dam type, such as gravity dams, arch dams, or embankment dams, depending on the specific site conditions. The construction process involves meticulous planning, site preparation, and the use of high-quality construction materials to ensure the structural integrity and longevity of the dam.

Reservoir Management:

Reservoirs act as water storage facilities, providing a reliable and consistent water supply for hydroelectric power plants. Efficient reservoir management is crucial to optimize power generation. By controlling the water release rates, the reservoir can regulate the flow of water to the turbines, ensuring a steady and constant energy output. Additionally, reservoirs can provide a range of secondary benefits, including irrigation, flood control, and recreational activities.

Environmental Considerations:

While dam and reservoir systems offer numerous advantages, they also impact the environment. The creation of reservoirs can alter ecosystems and habitats, leading to the displacement of wildlife and changes in water quality. However, modern hydroelectric engineering focuses on minimizing these environmental impacts through careful planning, fish passage systems, and ecological restoration efforts.

Conclusion:

Dam and reservoir systems are integral components of hydroelectric power generation, offering a sustainable and reliable source of renewable energy. Students in the field of energy engineering must understand the functionality and design principles of these systems to contribute effectively to the development and maintenance of hydroelectric power plants. By harnessing the power of water, we can continue to meet the world's energy needs while minimizing our environmental footprint.

Intake Structures and Water Conveyance Systems

In the world of hydroelectric engineering, intake structures and water conveyance systems play a crucial role in efficiently harnessing the power of water. These components form the initial stage of any hydroelectric project, ensuring a steady and reliable flow of water to the turbines for power generation.

An intake structure serves as the entrance point for water into the hydroelectric system. It is designed to control the flow of water, prevent debris from entering, and ensure the safety of aquatic life. There are various types of intake structures, including gravity intakes, diversion intakes, and submerged intakes. Each type has its own advantages and is selected based on factors such as the availability of water sources and environmental considerations.

Water conveyance systems are the networks of canals, pipes, and channels that transport water from the intake structure to the turbines. These systems are designed to minimize energy losses due to friction and elevation changes, ensuring that the maximum amount of water reaches the turbines with optimal pressure and velocity. Engineers employ various hydraulic principles to design efficient water conveyance systems, including the Bernoulli's principle, which relates the velocity and pressure of fluid flow.

In the field of energy engineering, students must understand the importance of designing intake structures and water conveyance systems that are tailored to specific site conditions. Factors such as the available water supply, topography, and environmental impacts must be carefully considered during the design phase. Additionally, students should be aware of the challenges associated with maintenance, such as the accumulation of sediment and the growth of aquatic plants, which can hinder the efficiency of the system if not properly managed.

The integration of advanced technologies, such as sensors and automation, has revolutionized the monitoring and control of intake structures and water conveyance systems. Students should familiarize themselves with these technologies and their applications to ensure the sustainable and efficient operation of hydroelectric power plants.

In conclusion, intake structures and water conveyance systems are vital components of hydroelectric engineering. Understanding their design principles, challenges, and the latest technological advancements is essential for students in the field of energy engineering. By mastering these concepts, future engineers can contribute to the development of sustainable and efficient hydroelectric power plants, ensuring a reliable and clean source of energy for generations to come.

Powerhouse and Generating Equipment

In the world of hydroelectric engineering, the powerhouse is the heart of a hydroelectric power plant. It is where the mechanical energy of flowing water is converted into electrical energy, providing sustainable and clean power to communities around the world. Understanding the powerhouse and the generating equipment within it is crucial for students studying energy engineering, as it forms the foundation of hydroelectric power generation.

The powerhouse is typically located at the base of a dam, where the water from the reservoir is released to flow through massive turbines. These turbines are the primary generating equipment and are responsible for converting the kinetic energy of the water into mechanical energy. There are different types of turbines used in hydroelectric power plants, including Francis, Kaplan, and Pelton turbines, each designed for specific conditions and water flow rates.

Once the mechanical energy is generated, it is transferred to the generator, another crucial component of the powerhouse. The generator consists of a rotor and a stator, which work together to convert the mechanical energy into electrical energy. The rotating motion of the turbine spins the rotor within the generator, creating a magnetic field that induces electrical current in the stator's windings. This current is then collected and transmitted through electrical systems to power infrastructure and end-users.

In addition to turbines and generators, the powerhouse also houses other important equipment, such as transformers, circuit breakers, and control panels. Transformers are used to increase the voltage of the generated electricity for efficient transmission over long distances, while circuit breakers protect the electrical system from overloads and faults. Control panels enable operators to monitor and control the entire power plant, ensuring its safe and efficient operation.

Students studying energy engineering should familiarize themselves with the various components of a hydroelectric powerhouse and understand their functions. They should also learn about the design considerations, such as water flow rates, head height, and environmental impacts, that influence the selection and sizing of these components. By gaining a comprehensive understanding of powerhouse and generating equipment, students can contribute to the development of sustainable and efficient hydroelectric power plants that harness the immense power of water to meet the world's energy needs.

In conclusion, the powerhouse and generating equipment play a vital role in hydroelectric power generation. It is through these components that the mechanical energy of flowing water is converted into electrical energy. Students studying energy engineering should grasp the fundamentals of the powerhouse and its equipment, including turbines, generators, transformers, circuit breakers, and control panels. By understanding their functions and design considerations, students can contribute to the advancement of hydroelectric engineering and the development of clean and sustainable energy solutions.

Transmission and Distribution Systems

In the realm of energy engineering, transmission and distribution systems are crucial components of harnessing water power and delivering it to end-users. These systems play a vital role in ensuring the efficient and reliable supply of electricity generated from hydroelectric sources. Understanding the intricacies of transmission and distribution systems is essential for students pursuing studies in energy engineering, as it forms the backbone of the power sector.

Transmission systems are responsible for carrying high-voltage electricity over long distances, connecting power plants to substations. These systems are designed to minimize energy losses during transmission, ensuring that the power reaches its destination with minimal dissipation. The transmission infrastructure consists of overhead transmission lines, underground cables, and transformers, among other components. Students will delve into the technicalities of transmission lines, understanding their different types, such as bundled conductors, and their varying voltage levels.

Distribution systems, on the other hand, focus on delivering electricity from substations to end-users, including residential, commercial, and industrial consumers. Students will explore the intricacies of distribution networks, which include transformers, circuit breakers, and distribution lines. They will learn about the various voltage levels used in distribution systems, such as low voltage (LV), medium voltage (MV), and high voltage (HV). Moreover, students will gain insight into the challenges faced by distribution systems, including voltage regulation, load management, and system protection.

It is essential for students to grasp the concepts of power flow, fault analysis, and protection schemes in transmission and distribution systems. They will learn about power system operations, including load dispatching, which involves managing the generation and distribution of electricity efficiently. Furthermore, students will be introduced to the concept of smart grids, which utilize advanced technologies to enhance the reliability, efficiency, and sustainability of transmission and distribution systems.

By studying transmission and distribution systems, students will gain a comprehensive understanding of the infrastructure and processes involved in delivering hydroelectric power to end-users. They will be equipped with the knowledge to design, analyze, and optimize these systems, ensuring a reliable and sustainable energy supply. With the global focus on renewable energy sources, hydroelectric engineering is becoming increasingly important, and students specializing in energy engineering will find themselves at the forefront of this field.

In conclusion, the subchapter on transmission and distribution systems in the book "Harnessing Water Power: A Comprehensive Introduction to Hydroelectric Engineering for Students" is an indispensable resource for students pursuing studies in energy engineering. It provides a comprehensive overview of the infrastructure, technicalities, and challenges associated with transmitting and distributing hydroelectric power. By delving into the intricacies of these systems, students will be well-prepared to contribute to the development of efficient and sustainable energy solutions in the field of hydroelectric engineering.

Chapter 6: Operation and Maintenance of Hydroelectric Power Plants

Operation Strategies and Control Systems

In the world of hydroelectric engineering, operation strategies and control systems are crucial aspects that ensure the efficient and optimal performance of hydroelectric power plants. This subchapter aims to provide students specializing in Energy Engineering with a comprehensive understanding of these essential concepts.

Operation strategies refer to the various techniques and approaches employed to manage the generation and distribution of hydroelectric power. These strategies are designed to maximize energy production while ensuring the long-term sustainability of the power plant. Students will explore the different factors and considerations involved in designing operation strategies, such as water availability, environmental impact, and electricity demand.

One key aspect of operation strategies is load balancing, which involves adjusting the power output of the hydroelectric plant to match the fluctuating electricity demand. Students will learn about the importance of load forecasting and how it helps in making informed decisions regarding power generation and distribution. They will also delve into the concept of peak shaving, which involves storing excess energy during low demand periods and releasing it during peak times, thus optimizing the overall efficiency of the power plant.

Control systems play a vital role in ensuring the safe and reliable operation of hydroelectric power plants. This subchapter will introduce students to the different types of control systems used in hydroelectric engineering, including supervisory control and data acquisition (SCADA) systems. They will gain insights into the functionalities and components of these systems, which enable operators to monitor and control various aspects of the power plant, such as water flow, turbine speed, and generator output.

Moreover, students will explore the importance of automation in control systems, as it enhances the efficiency and safety of hydroelectric power plants by reducing human error. They will learn about the integration of advanced technologies, such as artificial intelligence and machine learning, in control systems and their impact on the future of hydroelectric engineering.

Throughout this subchapter, students will have the opportunity to analyze real-world case studies and engage in practical exercises, allowing them to apply the concepts learned to solve operational challenges commonly encountered in hydroelectric power plants.

By the end of this subchapter, students will have a solid understanding of the operation strategies and control systems used in hydroelectric engineering. They will be equipped with the knowledge and skills necessary to optimize power generation, ensure the stability of the electrical grid, and contribute to sustainable energy solutions in the field of Energy Engineering.

Maintenance Procedures and Safety Measures

As students delving into the fascinating field of Energy Engineering, it is crucial to have a comprehensive understanding of maintenance procedures and safety measures in hydroelectric engineering. Hydroelectric power plants are complex systems that require regular maintenance to ensure optimal performance and safety. This subchapter aims to provide you with essential knowledge and guidelines to carry out maintenance procedures efficiently and prioritize safety.

Maintenance procedures in hydroelectric power plants encompass a wide range of activities, including routine inspections, preventive maintenance, and corrective maintenance. Regular inspections are vital for identifying potential issues before they escalate into more significant problems. These inspections should cover crucial components such as turbines, generators, transformers, and control systems. By adhering to a well-defined inspection schedule, you can detect early signs of wear, damage, or inefficiencies, allowing for timely repairs and reducing downtime.

Preventive maintenance is another critical aspect of ensuring the long-term performance of a hydroelectric power plant. This type of maintenance involves routine tasks such as lubrication, cleaning, and filter replacements. It also includes periodic testing and calibration of equipment to maintain their accuracy and efficiency. By implementing a preventive maintenance program, you can minimize unexpected breakdowns, optimize plant reliability, and extend the lifespan of equipment.

In addition to regular maintenance procedures, safety measures must be prioritized to prevent accidents and ensure the well-being of workers. Hydroelectric power plants pose several potential hazards, including electrical shocks, mechanical injuries, and falls. It is vital for all personnel to receive comprehensive training on safety protocols and to adhere to strict safety guidelines at all times. This includes the use of personal protective equipment (PPE), such as helmets, gloves, and safety harnesses, when working in potentially hazardous areas. Proper signage, barricades, and lockout/tagout procedures should also be implemented to prevent unauthorized access and ensure the safety of maintenance personnel.

Furthermore, it is essential to have emergency response plans in place to address unforeseen incidents effectively. These plans should include procedures for evacuations, first aid, and fire suppression. Regular drills and training sessions should be conducted to familiarize all personnel with these protocols, ensuring a swift and coordinated response in case of an emergency.

By understanding and implementing effective maintenance procedures and safety measures, you will contribute to the smooth operation, reliability, and safety of hydroelectric power plants. These practices are vital for energy engineers to master, as they ensure the sustainable utilization of water resources and the continuous generation of clean and renewable energy.

Performance Monitoring and Efficiency Optimization

In the realm of energy engineering, ensuring the efficient utilization of resources is of paramount importance. When it comes to hydroelectric power generation, performance monitoring and efficiency optimization play a critical role in maximizing the output and sustainability of such systems. This subchapter delves into the significance of monitoring performance and the strategies employed to optimize efficiency in hydroelectric engineering.

Performance monitoring involves the continuous assessment and analysis of various parameters to evaluate the functioning of a hydroelectric power plant. Students studying energy engineering need to understand the importance of monitoring key performance indicators (KPIs) such as power output, turbine efficiency, and water flow rates. By closely tracking these metrics, engineers can identify any deviations from expected performance and take corrective actions promptly.

Efficiency optimization, on the other hand, focuses on improving the overall efficiency of the hydroelectric system. This subchapter will delve into the different techniques used to optimize the performance of turbines, generators, and other components of a hydroelectric power plant. Students will learn about the intricacies of turbine design, including the selection of the optimal turbine type and its impact on efficiency. Furthermore, they will explore the significance of proper maintenance and regular inspections to ensure optimal performance over the long term.

Moreover, this subchapter will shed light on the role of advanced technologies such as remote sensing, data analytics, and artificial intelligence in performance monitoring and efficiency optimization.

Students will gain an understanding of how these cutting-edge tools can be harnessed to collect and analyze data in real-time, facilitating the detection of potential issues and the implementation of proactive measures.

To enhance the learning experience, case studies and practical examples will be included throughout this subchapter. These examples will highlight the challenges faced by energy engineers in optimizing the performance of hydroelectric power plants and the innovative solutions they employ. Additionally, the subchapter will provide guidance on the necessary steps to develop an efficient monitoring system, including the selection of appropriate sensors and data collection methodologies.

By delving into the intricacies of performance monitoring and efficiency optimization, this subchapter aims to equip students of energy engineering with the knowledge and skills necessary to excel in the field of hydroelectric engineering. Through the understanding and application of these principles, students will contribute to the sustainable and efficient utilization of water power, ensuring a greener and more sustainable future for generations to come.

Chapter 7: Environmental and Social Considerations in Hydroelectric Engineering

Environmental Impact Assessments

In the field of hydroelectric engineering, it is crucial to consider the environmental impact of harnessing water power. Environmental Impact Assessments (EIAs) play a vital role in ensuring sustainable development and minimizing negative environmental consequences. This subchapter will provide an overview of EIAs and their significance in the energy engineering niche.

EIAs are comprehensive evaluations conducted to identify and assess the potential environmental effects of a proposed hydroelectric project. These assessments are performed before project implementation to inform decision-making, ensure compliance with environmental regulations, and promote sustainable practices. By conducting an EIA, engineers and developers can evaluate the potential impacts on water quality, aquatic and terrestrial ecosystems, air quality, noise levels, and socio-economic aspects.

One of the primary objectives of an EIA is to identify potential environmental risks and propose mitigation measures. For instance, the assessment may identify potential impacts on fish populations due to changes in water flow and temperature. In such cases, engineers can propose measures such as fish ladders or bypass systems to allow for fish migration and minimize the disruption to aquatic ecosystems.

Additionally, EIAs also consider the socio-economic impacts of hydroelectric projects. These assessments evaluate the potential effects on local communities, including changes in employment, land use, and cultural heritage. By involving stakeholders and local communities in the EIA process, engineers can better understand their concerns and incorporate their perspectives into project planning and decision-making.

Furthermore, EIAs also play a crucial role in meeting regulatory requirements. Environmental agencies often require developers to submit an EIA as part of the permitting process. These assessments help ensure that projects comply with environmental laws and regulations, protecting both the natural environment and the interests of local communities.

In the energy engineering niche, understanding and conducting EIAs is essential for sustainable hydroelectric development. Students studying energy engineering need to grasp the importance of environmental impact assessments to design and implement projects that are environmentally responsible and socially acceptable.

In conclusion, Environmental Impact Assessments are fundamental tools in hydroelectric engineering that help evaluate and minimize the potential environmental and socio-economic impacts of harnessing water power. By conducting thorough EIAs, energy engineers can ensure the sustainable development of hydroelectric projects, meeting regulatory requirements, and addressing the concerns of local communities. Understanding the significance of EIAs is crucial for students in the energy engineering niche, enabling them to contribute to the development of environmentally sustainable energy solutions.

Mitigation of Environmental and Social Impacts

In the field of hydroelectric engineering, it is crucial to address the potential environmental and social impacts associated with harnessing water power. As students in the field of energy engineering, it is essential to develop a comprehensive understanding of these impacts and explore effective strategies for their mitigation. This subchapter aims to shed light on the various environmental and social challenges associated with hydroelectric projects and the mitigation measures that can be implemented.

One of the primary environmental impacts of hydroelectric projects is the alteration of natural river ecosystems. Dams and reservoirs can disrupt the flow of water, leading to changes in water temperature, sedimentation, and fish migration patterns. These alterations can have significant ecological consequences, affecting aquatic biodiversity and the overall health of river ecosystems. To mitigate these impacts, engineers can implement measures such as fish ladders and fish bypass systems to facilitate fish migration and maintain ecological balance.

Another critical aspect to consider is the displacement of local communities due to the construction of hydroelectric projects. Reservoirs created by dams can result in the submergence of land and the relocation of communities. It is essential to engage with the affected communities, ensuring their participation in decision-making processes and providing adequate compensation and resettlement options. Additionally, engineers can explore alternative designs that minimize the need for large-scale resettlement.

Furthermore, the construction and operation of hydroelectric projects can have socio-economic impacts on local communities. These can include changes in land use, employment opportunities, and cultural heritage. To address these impacts, engineers should conduct detailed social impact assessments, considering the needs and aspirations of the local population. Implementing sustainable development programs, such as the creation of alternative livelihood opportunities and investments in education and healthcare, can help mitigate the negative social effects.

In conclusion, the mitigation of environmental and social impacts is a crucial aspect of hydroelectric engineering. As students in the field of energy engineering, it is essential to be aware of these impacts and explore strategies to address them. By implementing measures such as fish ladders, engaging with affected communities, and conducting social impact assessments, engineers can contribute to the sustainable development of hydroelectric projects. This subchapter serves as a foundation for understanding these impacts and provides a starting point for further exploration and research in the field.

Sustainable Practices in Hydroelectric Power Generation

In recent years, there has been a growing concern about the environmental impact of traditional energy sources, leading to a strong push for sustainable alternatives. Hydroelectric power generation stands out as one of the most environmentally friendly options available today. This subchapter aims to provide students in the field of energy engineering with an overview of sustainable practices in hydroelectric power generation, highlighting its benefits and the challenges it faces.

One of the key advantages of hydroelectric power is its renewable nature. Unlike fossil fuels, which are finite resources and contribute to greenhouse gas emissions, hydroelectric power relies on the continuous flow of water, making it a sustainable and clean source of energy. By harnessing the power of flowing water, hydroelectric plants generate electricity without producing harmful pollutants or greenhouse gases, thus reducing the carbon footprint associated with power generation.

Another important aspect of sustainable hydroelectric power generation is its potential for energy storage. Large-scale hydroelectric projects often include reservoirs that store water, allowing for the production of electricity on-demand. This flexibility in power generation can help balance the intermittent nature of renewable energy sources like wind and solar, ensuring a stable and reliable electricity supply.

However, it is crucial to consider the potential environmental impacts of hydroelectric power generation. The construction of dams and reservoirs can alter ecosystems and affect wildlife habitats. Additionally, the creation of reservoirs can cause the accumulation of sediment, reducing the storage capacity and lifespan of the dam. It is important for engineers to carefully assess and mitigate these impacts through proper planning, design, and management of hydroelectric projects.

To enhance sustainability, modern hydroelectric projects incorporate innovative technologies. For instance, fish ladders are designed to help fish migrate upstream, mitigating the disruption caused by dams. Turbine designs are continually improving to reduce the impact on aquatic life and improve energy efficiency. Furthermore, advanced monitoring systems enable real-time data collection, allowing operators to optimize power generation while minimizing environmental impacts.

In conclusion, sustainable practices in hydroelectric power generation offer a promising solution to our energy needs while minimizing harm to the environment. By harnessing the power of flowing water, hydroelectric plants provide a renewable source of electricity, reducing greenhouse gas emissions and supporting a transition to cleaner energy systems. However, it is essential for energy engineers to consider the potential environmental impacts and incorporate innovative technologies to ensure the sustainability of hydroelectric projects. With ongoing research and development, hydroelectric power generation can continue to evolve, contributing significantly to a more sustainable future.

Chapter 8: Challenges and Future Trends in Hydroelectric Engineering

Addressing Climate Change and Water Availability Issues

In recent years, the global community has been facing a pressing challenge – climate change. This phenomenon has had far-reaching consequences, including the availability and distribution of water resources. As students specializing in Energy Engineering, it is crucial to understand the impact of climate change on water availability and how it affects hydroelectric engineering.

Climate change has altered weather patterns and intensified extreme weather events such as droughts and floods. These changes have significant implications for water availability, as they directly affect the hydrological cycle. Rising temperatures lead to increased evaporation rates, which can result in reduced water levels in rivers, lakes, and reservoirs. Consequently, this poses a threat to the generation of hydroelectric power.

To address these challenges, hydroelectric engineers need to adopt sustainable practices that prioritize water conservation and efficient utilization. One approach is to implement advanced water management techniques, such as cloud seeding and rainwater harvesting, to augment water supplies. These innovative methods can help mitigate the effects of reduced rainfall and ensure a steady flow of water for hydroelectric power generation.

Additionally, it is vital to integrate climate data and modeling into the planning and design stages of hydroelectric projects. By incorporating climate change projections, engineers can anticipate changes in water availability and design infrastructure that is resilient to future climatic conditions. This proactive approach ensures that hydroelectric projects can adapt to the evolving needs of a changing climate.

Furthermore, students studying Energy Engineering must also be aware of the importance of renewable energy sources in combating climate change. Hydroelectric power is a clean and renewable energy resource that can help reduce greenhouse gas emissions. By relying on hydroelectricity, we can minimize our dependence on fossil fuels and contribute to a more sustainable future.

In conclusion, the impact of climate change on water availability poses significant challenges for hydroelectric engineering. As students specializing in Energy Engineering, understanding these challenges is crucial for developing sustainable solutions. By adopting innovative water management techniques, integrating climate data into project planning, and promoting the use of renewable energy, we can address climate change and water availability issues effectively. Through these efforts, we can harness the power of water sustainably and contribute to a greener and more resilient future.

Integration of Hydroelectric Power with Other Renewable Energy Sources

In recent years, there has been a growing recognition of the need to transition from fossil fuels to more sustainable and renewable energy sources. As students in the field of Energy Engineering, it is crucial for us to understand the various ways in which different renewable energy sources can complement and integrate with each other. One such integration that holds immense potential is the combination of hydroelectric power with other renewable energy sources.

Hydroelectric power is already one of the most widely used and efficient sources of renewable energy. By harnessing the power of flowing water, hydroelectric plants can generate a significant amount of electricity with minimal environmental impact. However, one limitation of hydroelectric power is its dependency on water availability, which can vary seasonally or due to droughts. This is where the integration of other renewable energy sources can play a vital role.

Solar and wind energy are two prominent renewable energy sources that can be effectively integrated with hydroelectric power. Solar panels can be installed in the vicinity of hydroelectric plants, taking advantage of the ample open space often found in hydropower facilities. During peak sunlight hours, solar panels can generate electricity, which can be used to supplement the power generated by hydroelectric turbines. Similarly, wind turbines can be strategically placed near hydroelectric plants to capture wind energy and contribute to the overall electricity generation.

The integration of hydroelectric power with solar and wind energy offers several advantages. Firstly, it helps to diversify the energy mix, reducing the reliance on a single renewable energy source. This diversification enhances the overall reliability and stability of the power grid. Secondly, by combining different renewable energy sources, the intermittent nature of solar and wind energy can be mitigated. When solar or wind power generation decreases, the hydroelectric plant can compensate by increasing its output, ensuring a continuous supply of electricity.

Furthermore, the integration of hydroelectric power with other renewable sources can facilitate the storage of excess energy. During periods of high solar or wind power generation, the surplus electricity can be used to pump water into the reservoir of a hydroelectric plant, effectively storing the energy for later use. This pumped storage system acts as a large-scale energy storage solution, enabling a more balanced and flexible power supply.

In conclusion, the integration of hydroelectric power with other renewable energy sources holds immense potential in the field of Energy Engineering. By combining the reliable and efficient nature of hydroelectric power with the intermittent yet abundant solar and wind energy, a more sustainable and resilient energy system can be achieved. As students, it is crucial for us to explore and understand these integration strategies to pave the way for a greener and more sustainable future.

Advancements in Turbine Technology and Efficiency

In the dynamic field of hydroelectric engineering, turbine technology and efficiency play a crucial role in harnessing water power effectively and sustainably. Over the years, significant advancements have been made in this area, revolutionizing the way we generate clean and renewable energy. This subchapter explores the latest developments in turbine technology, aiming to provide students in the niche of energy engineering with a comprehensive understanding of the subject.

One of the major breakthroughs in turbine technology is the development of more efficient designs. Traditional turbines have given way to modern designs that are capable of extracting a higher percentage of energy from flowing water. These advanced turbines maximize energy conversion by utilizing innovative features such as adjustable blades, optimized shapes, and improved materials. By minimizing energy losses and increasing overall efficiency, these turbines contribute to higher power output and improved performance.

Furthermore, the advent of computational fluid dynamics (CFD) has revolutionized turbine design and analysis. CFD allows engineers to simulate and optimize the flow of water within turbines, enabling them to accurately predict performance and identify areas for improvement. This computational approach has significantly reduced the time and cost involved in turbine design, making it more accessible for engineers to experiment with different configurations and improve overall efficiency.

Moreover, advancements in materials science have played a pivotal role in turbine technology. New materials, such as advanced alloys and composites, offer enhanced durability and resistance to corrosion, enabling turbines to operate more reliably in harsh environments. These materials also contribute to reduced maintenance costs and increased turbine lifespan, making hydroelectric power a more economically viable option.

In addition to efficiency improvements, modern turbines also address environmental concerns. Fish-friendly turbine designs have been developed to minimize the impact on aquatic ecosystems. These turbines incorporate features such as fish-friendly screens, safe passageways, and low shear forces, ensuring the safe passage of fish and other aquatic organisms.

As students in the field of energy engineering, it is crucial to stay updated with the latest advancements in turbine technology. By understanding these advancements, you will be equipped to contribute to the development of more efficient and sustainable hydroelectric systems, paving the way for a greener future.

Chapter 9: Case Studies in Hydroelectric Engineering

Case Study 1: Large-Scale Hydroelectric Power Plant

Introduction:

In this case study, we will explore the fascinating world of large-scale hydroelectric power plants. These engineering marvels have played a vital role in meeting the increasing global demand for electricity while minimizing the environmental impact. As students of energy engineering, understanding the design and operation of such power plants is essential for a comprehensive understanding of hydroelectric engineering.

Background:

Large-scale hydroelectric power plants harness the energy of flowing water to generate electricity on a massive scale. They typically involve the construction of dams to create reservoirs, which store water. The stored water is then released through turbines, which convert the potential energy of the water into mechanical energy. This mechanical energy is further converted into electrical energy by generators, providing clean and renewable electricity to millions of people.

Design   and   Construction:

The design and construction of a large-scale hydroelectric power plant require meticulous planning and engineering expertise. Engineers must consider various factors, including the topography of the site, availability of water resources, and environmental impact assessments. The design also involves selecting the appropriate turbine and generator technologies, optimizing the dam's height and size, and ensuring the safety and stability of the entire structure.

Operation   and   Maintenance:
Once constructed, a large-scale hydroelectric power plant requires regular operation and maintenance to ensure optimal performance and longevity. Plant operators monitor the water level in the reservoir, control the release of water, and adjust the turbine and generator settings based on electricity demand. Maintenance activities include inspecting and repairing mechanical components, ensuring the integrity of the dam structure, and managing sediment accumulation in the reservoir.

Environmental   Impact:
While large-scale hydroelectric power plants provide clean and renewable energy, they can also have environmental consequences. The construction of dams may lead to the displacement of local communities and impact aquatic ecosystems. Additionally, altering the natural flow of rivers can affect downstream habitats and fish migration patterns. Engineers and environmentalists work together to mitigate these impacts through strategies like fish ladders, environmental flow releases, and habitat restoration initiatives.

Conclusion:
Large-scale hydroelectric power plants offer an efficient and sustainable solution to meet the world's growing energy needs. As students of energy engineering, understanding the design, operation, and environmental impact of these power plants is crucial. By exploring case studies like this, we gain valuable insights into the complex and dynamic field of hydroelectric engineering, paving the way for innovative solutions and a greener future.

Case Study 2: Small-Scale Hydroelectric Power Plant

Introduction:

In this case study, we will explore the concept of small-scale hydroelectric power plants and their significance in the field of energy engineering. Small-scale hydroelectric power plants have gained popularity due to their numerous advantages, including their renewable and sustainable nature, cost-effectiveness, and minimal environmental impact. This case study aims to provide a comprehensive understanding of the design, operation, and benefits of such power plants.

Design   and   Operation:

The design of a small-scale hydroelectric power plant involves harnessing the energy of flowing or falling water to generate electricity. It typically consists of a water source, such as a river or stream, a dam or diversion structure, a penstock, a turbine, a generator, and a transmission system. The water is collected and diverted through the penstock, which directs it towards the turbine. The force of the water turns the turbine, which in turn drives the generator to produce electricity. The electricity generated is then transmitted to the grid or used locally.

Benefits:

Small-scale hydroelectric power plants offer several advantages over other forms of energy generation. Firstly, they are a renewable and sustainable source of energy, as water is constantly replenished by rainfall or snowmelt. Secondly, these power plants have a relatively low operational cost and do not require expensive fuel. Thirdly, they have a long lifespan and require minimal maintenance, making them a reliable source of electricity for remote or off-grid areas. Moreover, small-scale hydroelectric power plants have a minimal environmental impact compared to fossil fuel power plants, as they do not emit greenhouse gases or contribute to air pollution.

Case Study Example:
Let's consider a case study of a small-scale hydroelectric power plant located in a rural mountainous region. The plant utilizes a nearby river with a consistent flow throughout the year. The design of this power plant includes a small dam to divert the water towards the penstock. The penstock transports the water downhill, and the force of the flowing water rotates a turbine, producing electricity. The generated electricity is distributed to nearby villages, providing them with a reliable and sustainable source of power.

Conclusion:
Small-scale hydroelectric power plants offer a promising solution for meeting the energy needs of remote areas while minimizing environmental impact. As students in the field of energy engineering, understanding the design and operation of such power plants is crucial. By harnessing the power of water, we can contribute to a greener and more sustainable future.

Case Study 3: Run-of-River Hydroelectric Power Plant

Introduction:
Welcome to Case Study 3 of our book "Harnessing Water Power: A Comprehensive Introduction to Hydroelectric Engineering for Students." In this subchapter, we will explore the fascinating world of run-of-river hydroelectric power plants. Specifically, we will delve into the design, operation, and benefits of these power plants, highlighting their significance in the field of energy engineering.

Overview of Run-of-River Hydroelectric Power Plants: Run-of-river hydroelectric power plants are innovative systems that harness the kinetic energy of flowing water to generate electricity. Unlike traditional hydroelectric power plants, run-of-river plants do not require large reservoirs. Instead, they utilize the natural flow of a river or stream to power turbines, resulting in a sustainable and eco-friendly energy source.

Design and Operation:
The design of run-of-river power plants involves diverting a portion of the river's flow through a channel or a penstock, directing the water to turbines. These turbines are connected to generators, which convert the mechanical energy of the flowing water into electrical energy. The generated electricity is then transmitted through power lines for distribution.

Benefits of Run-of-River Hydroelectric Power Plants:
1. Environmental Sustainability: Run-of-river power plants have minimal impact on the environment. Since they do not require large reservoirs, they avoid the negative consequences associated with damming rivers, such as flooding and displacement of wildlife. This

makes them an attractive choice for sustainable and responsible energy generation.

2. Renewable Energy Source: As run-of-river power plants utilize the natural flow of water, they have a virtually unlimited supply of renewable energy. Rivers and streams are constantly replenished by rainfall and snowmelt, ensuring a consistent source of clean power.

3. Cost Efficiency: Run-of-river power plants have lower construction and maintenance costs compared to traditional hydroelectric power plants. The absence of large reservoirs and dams reduces the investment required, making them a financially viable option for energy engineers.

4. Community Benefits: Run-of-river power plants have the potential to bring economic prosperity to local communities. They create job opportunities during the construction and operation phases, and the revenue generated from selling electricity can be reinvested in community development projects.

Conclusion:
Run-of-river hydroelectric power plants offer a sustainable and economical solution to energy engineering. With their environmentally friendly design, renewable energy source, cost efficiency, and community benefits, they represent a promising future for the energy sector. By understanding the principles and mechanics behind run-of-river power plants, students can contribute to the development of cleaner and more efficient energy systems, helping to create a greener and more sustainable world.

Chapter 10: Conclusion

Recap of Key Concepts

As we near the end of our journey through the world of hydroelectric engineering, it is crucial to recap the key concepts we have explored thus far. Understanding these fundamental principles will not only solidify your knowledge but also empower you to excel in the field of energy engineering. Let's take a moment to review the essential ideas we have covered.

1. Hydroelectric Power Generation: Hydroelectric power plants harness the energy of flowing water to generate electricity. They consist of various components, including dams, reservoirs, turbines, and generators. The potential energy of water stored in reservoirs is converted into kinetic energy, which drives the turbines and ultimately produces electrical power.

2. Types of Hydropower Systems: We have discussed different types of hydropower systems, such as impoundment, diversion, and pumped storage. Impoundment systems involve the construction of dams to store water in reservoirs. Diversion systems redirect water from a river through a canal or penstock to generate power. Pumped storage systems use excess electricity during low-demand periods to pump water uphill, which is then released to generate power during peak demand.

3. Environmental and Social Impact: While hydroelectric power offers numerous benefits, it is essential to consider its environmental and social impact. We have explored the potential effects on aquatic ecosystems, fish migration, and downstream flow. Additionally, the

displacement of communities due to dam construction and the alteration of natural landscapes must be carefully considered.

4. Efficiency and Optimization: We have learned about the factors that impact the efficiency of hydroelectric power generation, including head height, flow rate, and turbine design. Optimizing these factors can enhance the overall efficiency and performance of a hydroelectric power plant.

5. Future Trends: Finally, we have discussed emerging trends in hydroelectric engineering, such as the use of advanced turbine designs, fish-friendly technologies, and the integration of renewable energy sources. These advancements aim to make hydropower more sustainable, efficient, and environmentally friendly.

By revisiting these key concepts, you have solidified your understanding of hydroelectric engineering and its role in energy production. As aspiring energy engineers, it is crucial to continually expand your knowledge and stay updated with the latest developments in the field. Hydroelectric power is a vital component of the renewable energy sector, and your expertise in this area will be invaluable in shaping a sustainable future.

In the next chapter, we will delve into case studies of notable hydroelectric power plants worldwide, analyzing their design, construction, and operational challenges. This exploration will provide you with real-life examples and practical insights that further enhance your understanding of hydroelectric engineering.

Importance of Hydroelectric Engineering in the Future

The Importance of Hydroelectric Engineering in the Future

In recent years, the global energy landscape has been undergoing a significant transformation, driven by the need to find sustainable alternatives to fossil fuels. As the demand for clean and renewable energy sources continues to rise, the importance of hydroelectric engineering in the future cannot be overstated. This subchapter aims to shed light on the vital role that hydroelectric engineering plays in the field of energy engineering.

Hydroelectric power, generated by harnessing the force of flowing water, has long been recognized as one of the most reliable and efficient sources of renewable energy. Unlike other forms of renewable energy such as solar or wind power, hydroelectric power can provide a stable and consistent supply of electricity throughout the year. This makes it an ideal choice for meeting the growing energy demands of the future.

One of the key advantages of hydroelectric engineering lies in its ability to store large amounts of energy. By constructing dams and reservoirs, excess electricity can be generated during periods of low demand and stored for use during peak hours. This not only ensures a steady supply of electricity but also helps stabilize the grid, making hydroelectric power a crucial component of the future smart grid systems.

Furthermore, hydroelectric engineering offers a multitude of environmental benefits. Unlike fossil fuels, hydroelectric power does not emit harmful greenhouse gases, contributing to the mitigation of climate change. Additionally, hydroelectric power plants have the potential to provide water management solutions by regulating and controlling water flow. This plays a crucial role in flood control, irrigation, and maintaining ecological balance in river systems.

From an economic perspective, hydroelectric engineering projects have the potential to create numerous employment opportunities. The construction and maintenance of hydroelectric power plants require a diverse range of technical expertise, making it an attractive field for aspiring energy engineers. Moreover, the development of hydroelectric projects can stimulate economic growth in local communities, as it often involves infrastructure development and increased tourism.

In conclusion, the importance of hydroelectric engineering in the future cannot be emphasized enough. With its ability to provide a stable and reliable source of renewable energy, while also offering environmental and economic benefits, hydroelectric power is poised to play a significant role in meeting the energy demands of the future. Students pursuing a career in energy engineering should understand the principles and practices of hydroelectric engineering, as it will undoubtedly be a key area of focus in the years to come. By harnessing the power of water, we can make tremendous strides towards a sustainable and greener future.

Final Thoughts and Recommendations for Students

Congratulations! You have completed your journey through the world of hydroelectric engineering. As students in the field of energy engineering, you now possess a comprehensive understanding of the fascinating world of harnessing water power. Before we conclude, let us summarize the key takeaways from this book and provide you with some recommendations for your future endeavors.

Throughout this book, we have explored the principles, technologies, and applications of hydroelectric engineering. We have learned about the different types of hydroelectric power plants, including impoundment, diversion, pumped storage, and tidal plants. We have delved into the intricate mechanisms of dam construction, turbine design, and power generation. Moreover, we have examined the environmental impacts, sustainability, and future prospects of hydroelectric power.

As you embark on your journey as energy engineering students, it is crucial to remember the significance of hydroelectric power in the global energy landscape. Hydroelectric power is a clean, renewable, and reliable source of energy that plays a pivotal role in combating climate change and meeting the world's growing energy demands. By constantly striving for innovation and improvement, you can contribute to the development and optimization of hydroelectric power plants.

Here are some final thoughts and recommendations for you:

1. Embrace interdisciplinary learning: Hydroelectric engineering requires a multidisciplinary approach. Familiarize yourself with concepts from civil engineering, mechanical engineering,

environmental science, and sustainability to enhance your knowledge and skills.

2. Stay updated with technological advancements: The field of hydroelectric engineering is constantly evolving. Keep yourself informed about the latest technologies, materials, and design methodologies to stay at the forefront of innovation.

3. Foster collaboration: Collaborate with fellow students, professors, and professionals in the field. Engage in discussions, attend conferences, and participate in research projects to broaden your horizons and exchange ideas.

4. Prioritize sustainability: As energy engineers, it is our responsibility to ensure that hydroelectric projects are developed sustainably. Consider the environmental, social, and economic impacts of these projects, and strive for a balance that benefits both society and the environment.

5. Pursue further education and research: Consider pursuing advanced degrees or engaging in research opportunities to deepen your understanding and contribute to the field of hydroelectric engineering. There are numerous scholarships, grants, and fellowships available for aspiring researchers.

In conclusion, harnessing water power through hydroelectric engineering is a vital component of the global energy mix. As students in the field of energy engineering, you have the power to shape the future of sustainable energy. By applying the knowledge gained from this book and following the recommendations provided, you can make a significant impact in the world of hydroelectric engineering. Good luck on your journey!

www.ingramcontent.com/pod-product-compliance
Lightning Source LLC
Chambersburg PA
CBHW051350150726
48000CB00003B/1129